NAVTEX MANUAL

2017 EDITION

INTERNATIONAL MARITIME ORGANIZATION
London, 2017

First published in 1988
by the INTERNATIONAL MARITIME ORGANIZATION
4 Albert Embankment, London SE1 7SR
www.imo.org

Sixth edition 2017

Printed in the United Kingdom by CPI Colour

ISBN 978-92-801-1662-5

IMO PUBLICATION
Sales number: IE951E

Contents

Foreword		1
1	General information	3
2	NAVTEX service	3
3	General features of the NAVTEX system	10
4	Planning NAVTEX service	10
5	NAVTEX message technical characters	13
6	Message identity	16
7	Message format	17
8	Language and national broadcast options	21
9	Information control	21
10	Message content	22
11	Message priorities and broadcast procedures in the international NAVTEX service	23
12	Responsibilities of a NAVTEX Coordinator	24
13	Best practice for those using the service	26
14	Mutual interference between NAVTEX stations	27
15	Notification of NAVTEX services	28
Annex 1	IMO NAVTEX Coordinating Panel terms of reference	29
Annex 2	Recommendation ITU-R M.540-2 – Operational and technical characteristics for an automated direct-printing telegraph system for promulgation of navigational and meteorological warnings and urgent information to ships	31
Annex 3	Resolution MSC.148(77) – Adoption of the revised performance standards for narrow-band direct-printing telegraph equipment for the reception of navigational and meteorological warnings and urgent information to ships (NAVTEX)	37
Annex 4	Resolution A.801(19), annex 4 – Criteria for use when providing a NAVTEX service	41
Annex 5	Procedure for amending the NAVTEX Manual	43

Foreword

SOLAS regulation IV/12.2 states that "Every ship, while at sea, shall maintain a radio watch for broadcasts of maritime safety information on the appropriate frequency or frequencies on which such information is broadcast for the area in which the ship is navigating".

At the request of the IMO Sub-Committee on Radiocommunications (COM), the NAVTEX Manual was first produced in 1988. Three subsequent editions have been produced, with the fourth edition published in 2005 containing amendments endorsed by the Maritime Safety Committee (MSC) at its seventy-eighth session in May 2004 by MSC/Circ.1122.

At its seventh meeting in September 2005, the IHO Commission on the Promulgation of Radio Navigational Warnings (CPRNW)* established a working group to review all World-Wide Navigational Warning Service (WWNWS) documentation. The working group included representation from the WMO and firstly prepared revisions to Promulgation of Maritime Safety Information (resolutions A.705(17)) and A.706(17), World-Wide Navigational Warning Service. The proposed revisions of these resolutions were circulated to IHO Member States under IHB CL 104/2007, endorsed by the Sub-Committee on Radiocommunications and Search and Rescue (COMSAR) at its twelfth session in April 2008 and subsequently approved by the MSC at its eighty-fifth session in November/December 2008 by means of MSC.1/Circ.1287 and MSC.1/Circ.1288, respectively.

The IHO CPRNW Working Group then prepared the revised *Joint IMO/IHO/WMO Manual on Maritime Safety Information* incorporating the revised information from resolutions A.705(17), as amended, and A.706(17), as amended. The revised text was circulated to IHO Member States under cover of IHB CL 70/2008, endorsed by the COMSAR Sub-Committee at its thirteenth session in January 2009 and subsequently approved by the MSC at its eighty-sixth session in May/June 2009 by means of MSC.1/Circ.1310. The working group subsequently prepared the third revision of the International SafetyNET Manual. The revised text of the International SafetyNET Manual was circulated to IHO Member States under cover of IHB CL 68/2009, endorsed by the COMSAR Sub-Committee at its fourteenth session in March 2010 and approved by the MSC at its eighty-seventh session in May 2010 by MSC.1/Circ.1364.

Continuing with the holistic approach of reviewing all maritime safety information documents from the top-down, the working group prepared the fifth revision of the NAVTEX Manual. The revised text of the NAVTEX Manual was circulated to IHO Member States under cover of IHB CL 74/2010, endorsed by the COMSAR Sub Committee at its fifteenth session in March 2011 and subsequently approved by the MSC at its eighty-ninth session in May 2011 by MSC.1/Circ.1403.

Following the review of all WWNWS documentation, an editorial review has been conducted. As part of this editorial review, MSC.1/Circ.1287/Rev.1 and MSC.1/Circ.1288/Rev.1 were approved by the MSC at its ninety second session in June 2013, and MSC.1/Circ.1310/Rev.1 was approved by the MSC at its ninety-fourth session in November 2014. Following the approval of these circulars, the IHO WWNWS Sub-Committee Working Group reviewed the text of the NAVTEX Manual. This sixth revision of the NAVTEX Manual was endorsed by the IHO and the WMO through the WWNWS Sub-Committee, endorsed by the Sub-Committee on Navigation, Communications and Search and Rescue (NCSR) at its third session in February/March 2016 and subsequently approved by the MSC at its ninety-seventh session in November 2016 by means of MSC.1/Circ.1403/Rev.1, with an entry into force date of 1 January 2018.

* CPRNW was renamed the IHO WWNWS Sub Committee (WWNWS) with effect from 1 January 2009.

1 General information

1.1 NAVTEX is an international automated direct-printing service for promulgation of Maritime Safety Information (MSI), navigational and meteorological warnings, meteorological forecasts and other urgent safety-related messages to ships. It was developed to provide a low cost, simple and automated means of receiving MSI on board ships at sea in coastal waters. The information transmitted may be relevant to all sizes and types of vessel and the selective message-rejection feature ensures that mariners can receive MSI broadcasts which are tailored to their particular needs.

1.2 NAVTEX fulfils an integral role in the Global Maritime Distress and Safety System (GMDSS) developed by the International Maritime Organization (IMO) and incorporated into the 1988 amendments to the International Convention for the Safety of Life at Sea (SOLAS), 1974, as amended, as a requirement for ships to which the Convention applies.

1.3 This Manual describes the structure and operation of the NAVTEX service. It is intended primarily for use by national Administrations and others concerned with the preparation and broadcasting of MSI. It will also be of interest to seafarers, shipowners and others who need to receive such information in order to safely go about their business at sea. It should be used in conjunction with the *Joint IMO/IHO/WMO Manual on Maritime Safety Information* (also published as the *IHO/IMO World-Wide Navigational Warning Service Guidance Document*, IHO Publication S-53).

2 NAVTEX service

2.1 Introduction

2.1.1 NAVTEX provides shipping with navigational and meteorological warnings, meteorological forecasts and other urgent safety-related messages (as listed in table 1, section 5) by automatic display or printout from a dedicated receiver. It is suitable for use in all sizes and types of ships. Figure 1 illustrates the way the service is typically structured.

2.1.2 NAVTEX is a component of the IMO/IHO World-Wide Navigational Warning Service (WWNWS) defined by resolutions A.706(17), as amended, and A.1051(27), as amended. It has also been included as an element of the Global Maritime Distress and Safety System (GMDSS).

2.1.3 In the GMDSS, a NAVTEX receiving capability is part of the mandatory equipment which is required to be carried in certain vessels under the provisions of the International Convention for the Safety of Life at Sea (SOLAS), 1974, as amended.

2.1.4 Authority for coordinating the use of the frequencies 518 kHz, 490 kHz and 4209.5 kHz for NAVTEX services worldwide was delegated by ITU to IMO at WRC 95 through resolution 339. This was reaffirmed at WRC-97. IMO has vested responsibility for the overall management and coordination of the global NAVTEX service to the NAVTEX Coordinating Panel. With respect to National NAVTEX broadcasts on 490 kHz and 4209.5 kHz, the function of the NAVTEX Coordinating Panel is limited to the allocation of transmission identification characters.* It should be noted that the provisions of the NAVTEX Manual do not apply when planning a National NAVTEX service on other nationally assigned frequencies. The Terms of Reference for this Panel are attached in annex 1.

2.1.5 Details of operational and planned NAVTEX services are published periodically in the various national lists of radio signals, in an annex to the International Telecommunication Union's (ITU) List IV – List of coast stations and special service stations, and in the GMDSS Master Plan published by IMO in its series of GMDSS Circulars.

---

* The transmitter identification character is a single letter allocated to each transmitter to identify the NAVTEX station and broadcast times.

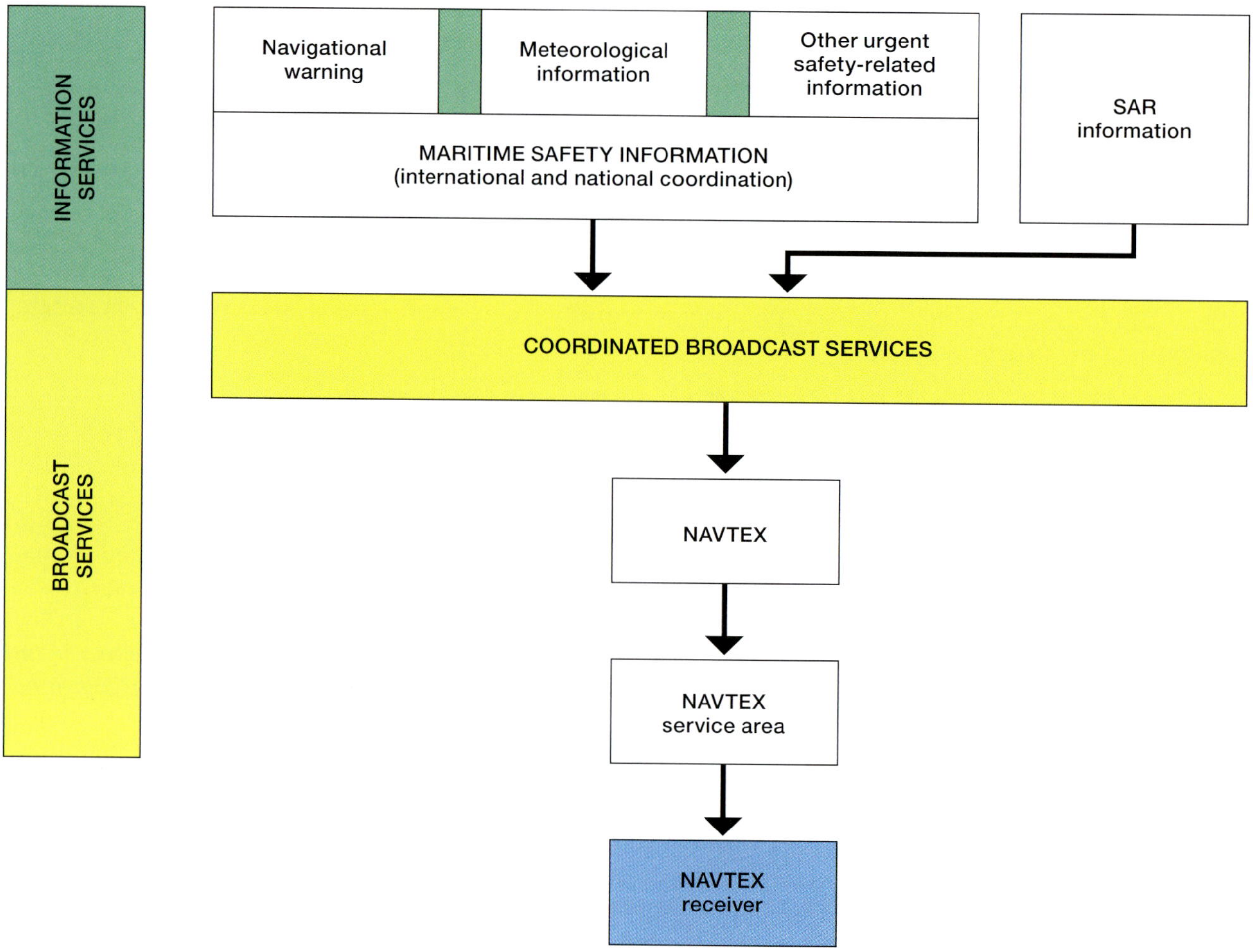

Figure 1 – *Basic concept of the NAVTEX system*

2.2 Definitions

2.2.1 For the purposes of this Manual, the following definitions apply:

.1 *Coastal warning* means a navigational warning or in-force bulletin promulgated as part of a numbered series by a National Coordinator. Broadcast should be made by the International NAVTEX service to defined NAVTEX service areas and/or by the International SafetyNET service to coastal warning areas. (In addition, Administrations may issue coastal warnings by other means).

.2 *Coastal warning area* means a unique and precisely defined sea area within a NAVAREA/METAREA or Sub-area established by a coastal State for the purpose of coordinating the broadcast of coastal maritime safety information through the SafetyNET service.

.3 *Global Maritime Distress and Safety System (GMDSS)* means the global communications service based upon automated systems, both satellite and terrestrial, to provide distress alerting and promulgation of maritime safety information for mariners.

.4 *HF NBDP* means High Frequency narrow-band direct-printing, using radio telegraphy as defined in Recommendation ITU-R M.688.

.5 *In-force bulletin* means a list of serial numbers of those NAVAREA, Sub area or coastal warnings in force issued and broadcast by the NAVAREA Coordinator, Sub-area Coordinator or National Coordinator.

.6 *International NAVTEX service* means the coordinated broadcast and automatic reception on 518 kHz of maritime safety information by means of narrow-band direct-printing telegraphy using the English language.*

.7 *International SafetyNET service* means the coordinated broadcast and automatic reception of maritime safety information via the Inmarsat Enhanced Group Call (EGC) system, using the English language, in accordance with the provisions of the International Convention for the Safety of Life at Sea, 1974, as amended.

.8 *Issuing Service* means a National Meteorological Service which has accepted responsibility for ensuring that meteorological warnings and forecasts for shipping are disseminated through the International SafetyNET service to the METAREA for which the Service has accepted responsibility under the broadcast requirements of the GMDSS.†

.9 *Local warning* means a navigational warning which covers inshore waters, often within the limits of jurisdiction of a harbour or port authority.

.10 *Maritime safety information (MSI)*‡ means navigational and meteorological warnings, meteorological forecasts and other urgent safety related messages broadcast to ships.

.11 *Maritime safety information service* means the internationally and nationally coordinated network of broadcasts containing information which is necessary for safe navigation.

.12 *METAREA* means a geographical sea area§ established for the purpose of coordinating the broadcast of marine meteorological information. The term METAREA followed by a roman numeral may be used to identify a particular sea area. The delimitation of such areas is not related to and shall not prejudice the delimitation of any boundaries between States (see figure 2).

.13 *METAREA Coordinator* means the authority charged with coordinating marine meteorological information broadcasts by one or more National Meteorological Services acting as Preparation or Issuing Services within the METAREA.

.14 *Meteorological information* means the marine meteorological warning and forecast information in accordance with the provisions of the International Convention for the Safety of Life at Sea, 1974, as amended.

.15 *National Coordinator* means the national authority charged with collating and issuing coastal warnings within a national area of responsibility.

.16 *National NAVTEX service* means the broadcast and automatic reception of maritime safety information by means of narrow-band direct printing telegraphy using frequencies other than 518 kHz and languages as decided by the Administration concerned.

.17 *National SafetyNET service* means the broadcast and automatic reception of maritime safety information via the Inmarsat EGC system, using languages as decided by the Administration concerned.

.18 *NAVAREA* means a geographical sea area§ established for the purpose of coordinating the broadcast of navigational warnings. The term NAVAREA followed by a roman numeral may be used to identify a particular sea area. The delimitation of such areas is not related to and shall not prejudice the delimitation of any boundaries between States (see figure 3).

.19 *NAVAREA Coordinator* means the authority charged with coordinating, collating and issuing NAVAREA warnings for a designated NAVAREA.

* As set out in this Manual.

† In the context of this Manual, *designated area* means the NAVTEX service area.

‡ As defined in regulation IV/2 of the 1974 SOLAS Convention, as amended.

§ Which may include inland seas, lakes and waterways navigable by seagoing ships.

.20 *NAVAREA warning* means a navigational warning or in-force bulletin promulgated as part of a numbered series by a NAVAREA Coordinator.

.21 *Navigational warning* means a message containing urgent information relevant to safe navigation broadcast to ships in accordance with the provisions of the International Convention for the Safety of Life at Sea, 1974, as amended.

.22 *NAVTEX* means the system for the broadcast and automatic reception of maritime safety information by means of narrow-band direct printing telegraphy.*

.23 *NAVTEX Coordinator* means the authority charged with operating and managing one or more NAVTEX stations broadcasting maritime safety information as part of the International NAVTEX service.

.24 *NAVTEX coverage area* means an area defined by an arc of a circle having a radius from the transmitter calculated according to the method and criteria given in resolution A.801(19), as amended, see annex 4.

.25 *NAVTEX service area* means a unique and precisely defined sea area, wholly contained within the NAVTEX coverage area, for which maritime safety information is provided from a particular NAVTEX transmitter. It is normally defined by a line that takes full account of local propagation conditions and the character and volume of information and maritime traffic patterns in the region, as given in resolution A.801(19), as amended, see annex 4.

.26 *Other urgent safety-related information* means maritime safety information broadcast to ships that is not defined as a navigational warning or meteorological information. This may include, but is not limited to, significant malfunctions or changes to maritime communications systems, and new or amended mandatory ship reporting systems or maritime regulations affecting ships at sea.

.27 *Rescue Coordination Centre (RCC)* means a unit responsible for promoting efficient organization of search and rescue services and for coordinating the conduct of search and rescue operations within a search and rescue region. **Note:** The term RCC will be used within this Manual to apply to either joint, aeronautical or maritime centres; JRCC, ARCC or MRCC will be used as the context warrants.

.28 *SafetyNET* means the international service for the broadcast and automatic reception of maritime safety information via the Inmarsat EGC system. SafetyNET receiving capability is part of the mandatory equipment which is required to be carried by certain ships in accordance with the provisions of the International Convention for the Safety of Life at Sea, 1974, as amended.

.29 *SAR information* means distress alert relays and other urgent search and rescue information broadcast to ships.

.30 *Sub-area* means a subdivision of a NAVAREA/METAREA in which a number of countries have established a coordinated system for the promulgation of maritime safety information. The delimitation of such areas is not related to and shall not prejudice the delimitation of any boundaries between States.

.31 *Sub-area Coordinator* means the authority charged with coordinating, collating and issuing Sub-area warnings for a designated Sub-area.

.32 *Sub-area warning* means a navigational warning or in-force bulletin promulgated as part of a numbered series by a Sub-area Coordinator. Broadcast should be made by the International NAVTEX service to defined NAVTEX service areas or by the International SafetyNET service (through the appropriate NAVAREA Coordinator).

.33 *UTC* means Coordinated Universal Time which is equivalent to GMT (or ZULU) as the international time standard.

* See annex 2.

.34 *World-Wide Met-ocean Information and Warning Service (WWMIWS)*[*] means the internationally coordinated service for the promulgation of meteorological warnings and forecasts.

.35 *World-Wide Navigational Warning Service (WWNWS)*[†] means the internationally and nationally coordinated service for the promulgation of navigational warnings.

.36 In the operating procedures, *coordination* means that the allocation of the time for data broadcast is centralized, the format and criteria of data transmissions are compliant as described in the *Joint IMO/IHO/WMO Manual on Maritime Safety Information* and that all services are managed as set out in resolutions A.705(17), as amended, A.706(17), as amended, and A.1051(27), as amended.

[*] As set out in resolution A.1051(27), as amended.

[†] As set out in resolution A.706(17), as amended.

2.2.2 *Delimitation of METAREAs*

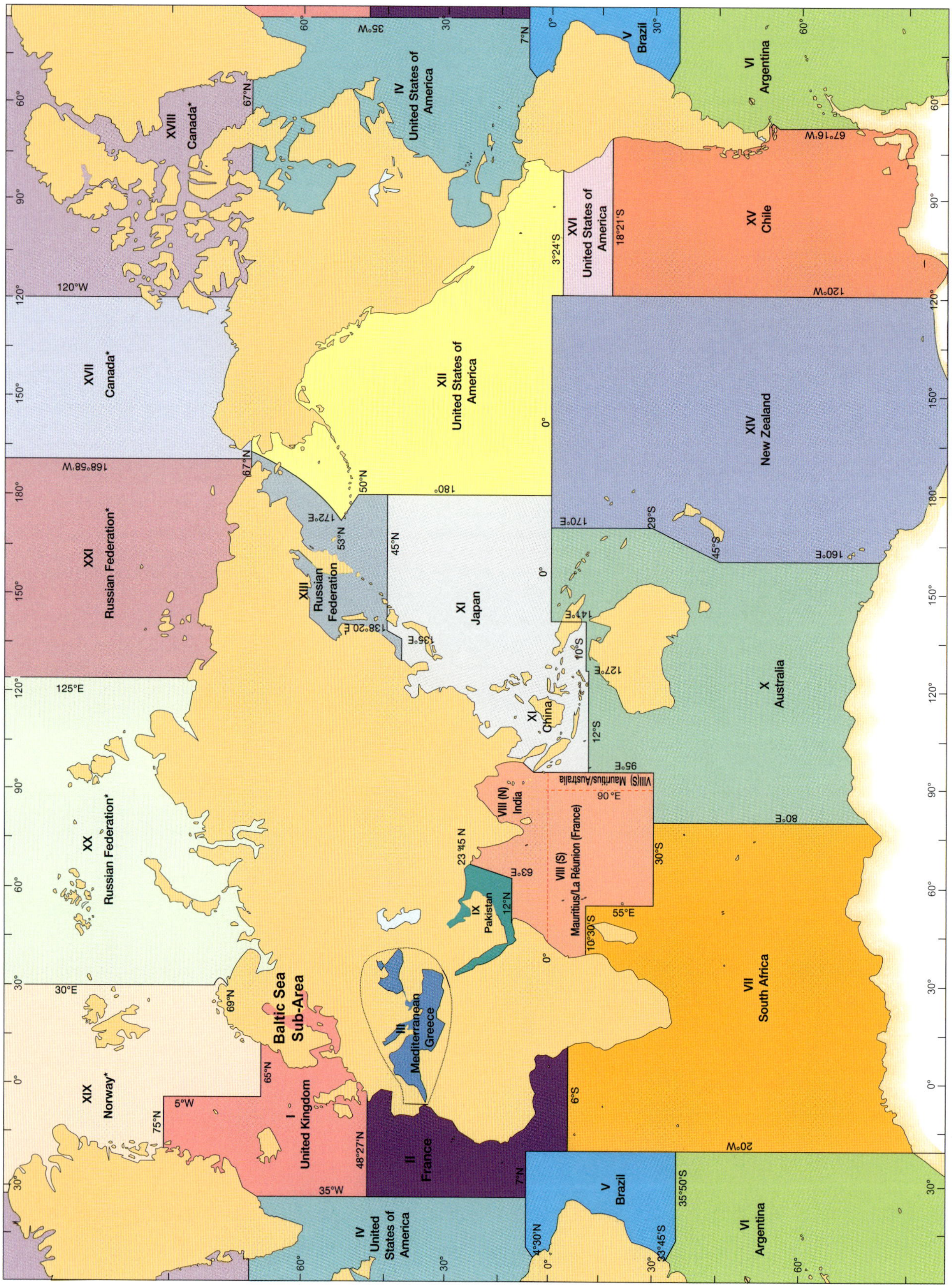

Figure 2 – METAREAs for coordinating and promulgating navigational warnings under the World-Wide Navigational Warning Service. The delimitation of such areas is not related to and shall not prejudice the delimitation of any boundaries between States.

2.2.3 *Delimitation of NAVAREAs*

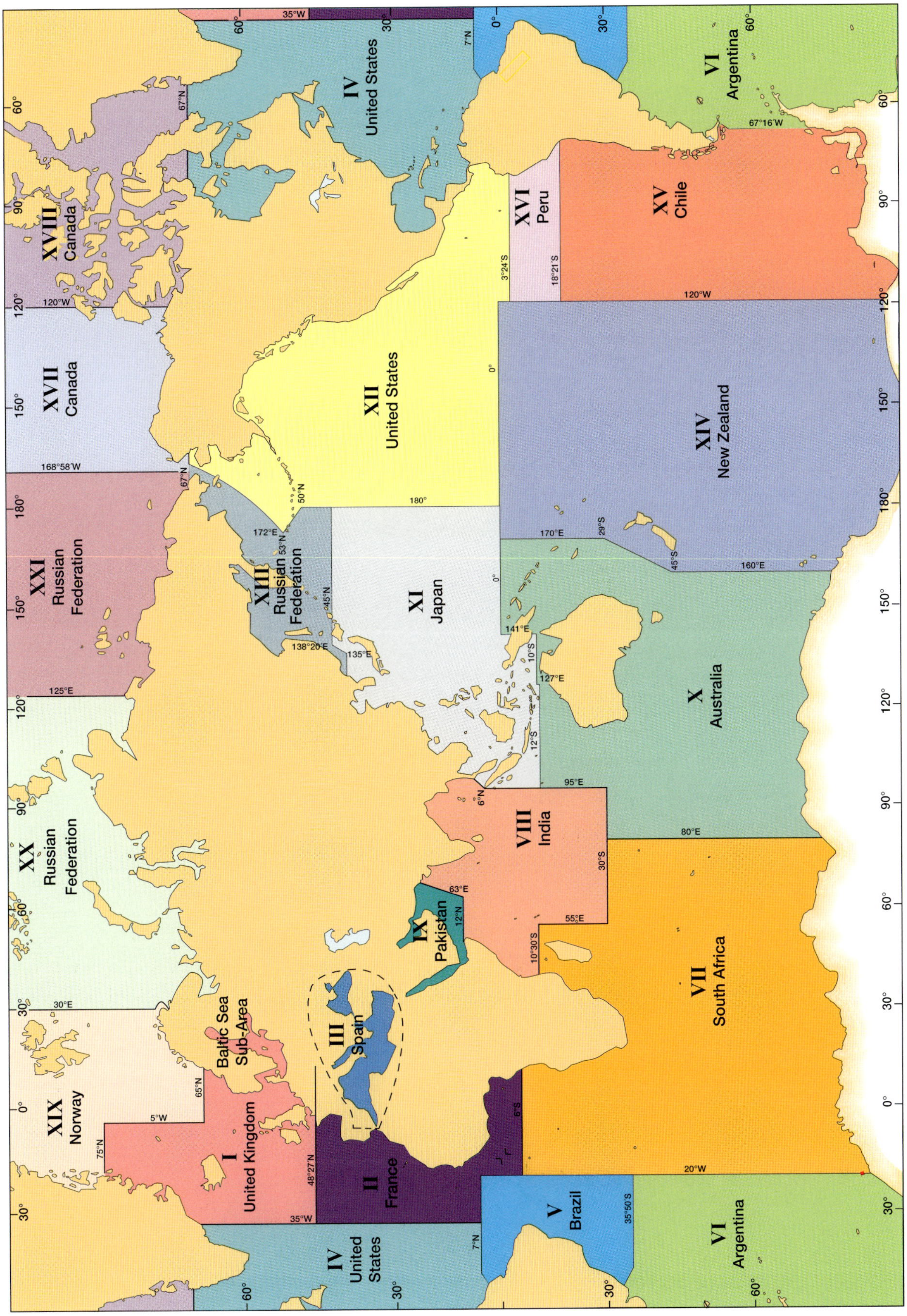

Figure 3 – *NAVAREAs for coordinating and promulgating navigational warnings and forecasts within the GMDSS. The delimitation of such areas is not related to and shall not prejudice the delimitation of any boundaries between States.*

3 General features of the NAVTEX system

3.1 The principal features are:

.1 use of a single frequency, with transmissions from stations within and between NAVAREAs and METAREAs coordinated on a time-sharing basis to reduce the risk of mutual interference. The following frequencies may be used for NAVTEX broadcasts:

518 kHz	
Type of service	International
Content	Maritime safety information
Language	English
Coordination	By IMO NAVTEX Coordinating Panel
490 kHz and 4209.5 kHz	
Type of service	National
Content	Maritime safety information
Language	As selected by the national Administration
Coordination	Transmitter identification character allocated by IMO NAVTEX Coordinating Panel
Other national frequencies allocated by the ITU	
Type of service	National
Content	As selected by the national Administration
Language	As selected by the national Administration
Coordination	By appropriate national Administration

.2 a dedicated NAVTEX receiver, comprising radio receivers, a signal processor and either:

.2.1 an integrated printing device; or

.2.2 a dedicated display device with a printer output port and a non-volatile message memory; or

.2.3 a connection to an integrated navigation system and a non-volatile message memory;

which has the ability to select messages to be printed, or viewed and stored in a memory according to:

– a technical code ($B_1B_2B_3B_4$), which appears in the preamble of each message; and

– whether or not the particular message has already been printed/received.

3.2 The operational and technical characteristics of the NAVTEX system are contained in Recommendation ITU-R M.540-2.[*] Performance standards for shipborne equipment, if installed before 1 July 2005, are laid down in resolution A.525(13). If installed on or after 1 July 2005, they shall conform to resolution MSC.148(77).[†]

4 Planning NAVTEX services

4.1 When planning NAVTEX services, Administrations should obtain guidance at an early stage from IMO, through its NAVTEX Coordinating Panel. This may be particularly important when installation of new stations and/or purchase of new equipment is under consideration. Details of how to contact the Panel may be found in annex 1.

[*] See annex 2.

[†] See annex 3.

4.2 International NAVTEX services on 518 kHz

When planning an International NAVTEX service it is essential to appreciate the high level of national and international coordination required. The central principles which should be borne in mind are as follows:

.1 All NAVTEX stations are part of the strategic infrastructure of both the GMDSS and WWNWS.

.2 It is essential for the efficiency and effectiveness of the service that a minimum number of stations are used. This may require national Administrations to either share facilities or promulgate information provided by Administrations of other nations.

.3 Each station contributes to the overall service in a coordinated way, bearing in mind the geographical area covered by each station and the effective coordination and control of information to be transmitted.

.4 The two basic areas which must be defined when establishing a NAVTEX station are the NAVTEX coverage area and the NAVTEX service area. Each station will provide all the information for a particular NAVTEX service area. The boundaries of the NAVTEX service area must be wholly contained within the coverage area, and must not overlap with adjacent NAVTEX service areas (see figure 4).

.5 National Administrations seeking to establish NAVTEX services shall undertake preliminary discussions with the NAVAREA Coordinator, METAREA Coordinator and neighbouring Administrations prior to formal application to IMO through the IMO NAVTEX Coordinating Panel. These discussions shall consider the most appropriate NAVTEX service area boundaries, possible geographical locations for transmitter sites to ensure optimal coverage and links with Information Providers.

.6 The range of a NAVTEX transmitter depends on the transmitted power and local radio propagation conditions. The actual range achieved shall be adjusted to the minimum required for adequate reception in the specified NAVTEX service area, taking into account the needs of ships approaching from other areas. Experience indicates that the required range of 250 to 400 nautical miles will normally be attained by transmitted power of no more than 1 kW during daylight **with a 60% reduction during night conditions**.

.7 After the choice of transmitter sites, the main need for coordination lies in the assignment of B_1 transmitter identification characters (time schedules) and the agreement of proposed NAVTEX service areas (if appropriate). Preliminary discussions between national Administrations seeking to establish or amend NAVTEX services and neighbouring Administrations shall be coordinated by the NAVAREA Coordinator prior to formal application for a B_1 transmitter identification character. Throughout the process the IMO NAVTEX Coordinating Panel is available to advise and liaise on the final limits of NAVTEX service areas if these cannot be agreed locally.

.8 The IMO NAVTEX Coordinating Panel will only allocate B_1 transmitter identification characters after the NAVTEX service areas have been agreed.

.9 Once a NAVTEX transmitter has been declared operational, if a national Administration wishes to:

.9.1 move the transmitter site; and/or

.9.2 amend the limits of its NAVTEX service area,

then the whole coordination process outlined above must be repeated, keeping the NAVTEX Coordinating Panel informed at all times.

.10 A National NAVTEX Coordinator shall be established to oversee the operation of the NAVTEX services established by each national Administration. The responsibilities of the NAVTEX Coordinator are defined in section 12 of this Manual.

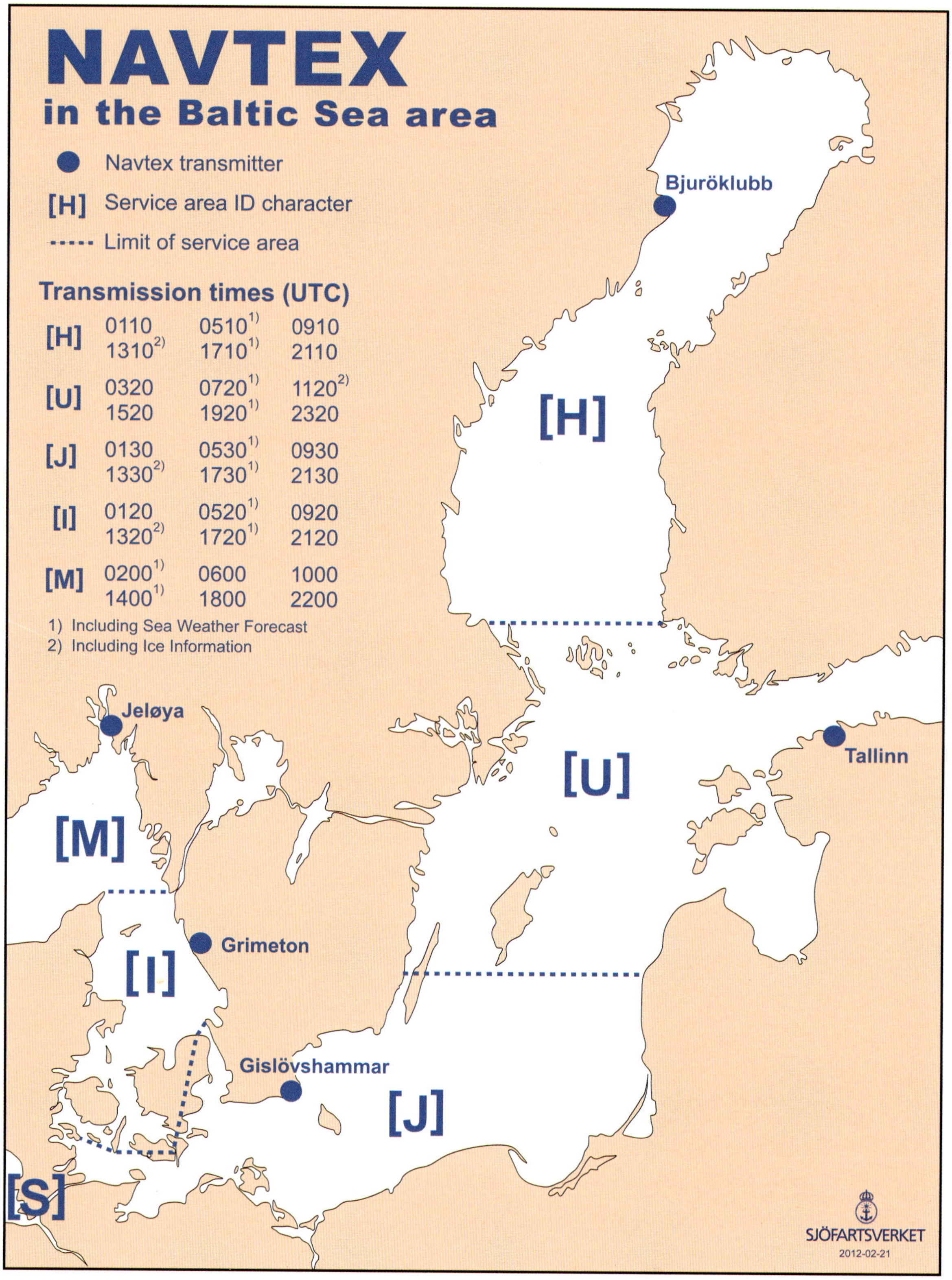

Figure 4 – *Example of NAVTEX service areas*

The Baltic Sea and its approaches have been divided into five individual NAVTEX service areas. Within each service area, maritime safety information is provided from a separate NAVTEX station which has been allocated a dedicated B_1 transmitter identification character. It is a fundamental requirement that the range of each NAVTEX transmitter is sufficient to include the whole of the NAVTEX service area assigned to its B_1 transmitter identification character.

4.3 National NAVTEX services on 490 kHz or 4209.5 kHz

The provisions of the NAVTEX Manual apply to National NAVTEX services on 490 kHz or 4209.5 kHz. When planning a National NAVTEX service, the IMO NAVTEX Coordinating Panel is responsible for the allocation of B_1 transmitter identification characters; however, the establishment of NAVTEX service areas and the compulsory use of the English language are not required.

4.4 National NAVTEX services on other frequencies

The provisions of the NAVTEX Manual do not apply when planning a National NAVTEX service on nationally assigned frequencies.

5 NAVTEX message technical characters

5.1 Overview of technical characters, B_1, B_2, B_3, B_4

5.1.1 NAVTEX messages include instructions to the NAVTEX receiver for processing maritime safety information in the form of the NAVTEX message identity, which consists of four technical "B" characters which make up an alphanumeric code. In order for messages to be correctly processed, they must consist of data conforming to these "B" characters:

B_1	Transmitter identification character
B_2	Subject indicator character
B_3B_4	Message numbering characters

5.2 B_1 – Transmitter identification character

5.2.1 The transmitter identification character (B_1) is a single letter which is allocated to each transmitter. It is used to identify the broadcasts which are to be accepted by the receiver and those to be rejected, and also the time slot for the transmission.

5.2.2 In order to avoid erroneous reception and interference of transmissions from two stations having the same transmitter identification character, it is necessary to ensure that such stations have a large geographical separation. Allocation of transmitter identification characters by alphabetical sequence to adjacent sites can also cause problems; hence, consecutive transmitter identification characters are not normally allocated to adjacent stations. Experience has shown that this removes the risk of a station which over-runs its time slot masking the phasing signal of an adjacent station which is about to begin its transmission.

5.2.3 NAVTEX transmissions have a designed maximum range of about 400 nautical miles. The minimum distance between two transmitters with the same transmitter identification identifier must, therefore, be sufficient to ensure that a receiver cannot be within range of both at the same time.

5.2.4 Close coordination between transmitting stations in adjacent NAVAREAs/METAREAs is necessary to achieve this separation. For this reason, national Administrations shall request the advice of the IMO NAVTEX Coordinating Panel at an early stage in the planning of a new NAVTEX service. The Panel will allocate B_1 transmitter identification characters in such a way as to minimize the risk of interference occurring.

5.2.5 Table 2 shows the transmitter identification characters and their associated transmission start times used by the IMO NAVTEX Coordinating Panel to evaluate and allocate transmitter identification characters A to X, regardless of the geographical position of the station anywhere in the world. Each transmitter identification character is allocated a maximum transmission time of 10 minutes every 4 h. **Because the NAVTEX system always utilizes a single frequency, it is fundamental to its successful operation that the following time slots are strictly adhered to, and that broadcasts do not overrun their allotted 10 minutes.**

5.2.6 In some regions, it has become necessary to accommodate a large number of stations. In extreme cases, it has even been necessary to reuse some transmitter identification characters for a second time within a region. Where this occurs every effort is made to ensure stations with the same character are as far apart as possible to reduce the risk of mutual interference.

Table 1 – *Technical "B" characters which make up the full NAVTEX message identity*

B_1 Transmitter identification character	B_2 Subject indicator character	B_3 B_4 Message numbering characters
1 letter	1 letter	2 digits
A to X	A = Navigational warnings	01 to 99 (message numbering characters "00" are not to be used for routine messages)
	B = Meteorological warnings	
	C = Ice reports	
	D* = Search and rescue information, acts of piracy warnings, tsunamis and other natural phenomena	
	E = Meteorological forecasts	
	F = Pilot and VTS service messages	
	G = AIS service messages (non-navigational aid)	
	H = LORAN messages	
	I = currently not used	
	J = GNSS messages	
	K = Other electronic navigational aid system messages	
	L = Other navigational warnings – additional to B_2 character A†	
	M =, N =, O =, P =, Q =, R =, S =, T =, U = } Currently not used	
	V =, W =, X =, Y = } Special services allocation by the IMO NAVTEX Coordinating Panel	
	Z = No message on hand	

* Use of B_2 character D will automatically set off the alarm at the NAVTEX receiver.

† On some older NAVTEX receivers it may be possible to deselect B_2 character L (continuation of B_2 subject group A), however, it is strongly recommended that this character is not deselected.

Table 2 – *NAVTEX transmission start times*

Transmitter identification character (B_1)	Transmission start times (UTC)					
A	0000	0400	0800	1200	1600	2000
B	0010	0410	0810	1210	1610	2010
C	0020	0420	0820	1220	1620	2020
D	0030	0430	0830	1230	1630	2030
E	0040	0440	0840	1240	1640	2040
F	0050	0450	0850	1250	1650	2050
G	0100	0500	0900	1300	1700	2100
H	0110	0510	0910	1310	1710	2110
I	0120	0520	0920	1320	1720	2120
J	0130	0530	0930	1330	1730	2130
K	0140	0540	0940	1340	1740	2140
L	0150	0550	0950	1350	1750	2150
M	0200	0600	1000	1400	1800	2200
N	0210	0610	1010	1410	1810	2210
O	0220	0620	1020	1420	1820	2220
P	0230	0630	1030	1430	1830	2230
Q	0240	0640	1040	1440	1840	2240
R	0250	0650	1050	1450	1850	2250
S	0300	0700	1100	1500	1900	2300
T	0310	0710	1110	1510	1910	2310
U	0320	0720	1120	1520	1920	2320
V	0330	0730	1130	1530	1930	2330
W	0340	0740	1140	1540	1940	2340
X	0350	0750	1150	1550	1950	2350

Figure 5 – *Example of NAVTEX receiver with LCD screen*

5.3 B_2 – Subject indicator character

5.3.1 Information is grouped by subject in the NAVTEX broadcast and each subject group is allocated a B_2 subject indicator character.

5.3.2 The subject indicator character is used by the receiver to identify the different classes of messages as listed in table 1.

5.3.3 Some subject indicator characters can be used to reject messages concerning certain subjects which may not be required by the ship (e.g. Ice report messages may be rejected by deselecting the B_2 subject indicator character C on the NAVTEX receiver on board a ship).

5.3.4 Reception of messages, transmitted using subject indicator characters A, B, D and L, which have been allocated for navigational warnings, meteorological warnings, search and rescue information, acts of piracy warnings, tsunamis and other natural phenomena, is mandatory and cannot be deselected on the NAVTEX receiver. This has been designed to ensure that ships using NAVTEX always receive the most essential information.

5.3.5 It is not possible to transmit or receive two NAVTEX messages with the same NAVTEX message identity (made up of the four technical characters). Therefore the B_2 subject indicator character L has been designated for use in the unlikely event that a NAVTEX Coordinator has more than 99 navigational warning messages in force and requiring transmission at the same time, all using B_2 subject indicator character A, with the same B_1 transmitter identification character.

5.3.6 Messages received which have been transmitted using subject indicator character D will set off an alarm built into the NAVTEX receiver.

5.3.7 In the International NAVTEX service, Administrations shall obtain the agreement of the IMO NAVTEX Coordinating Panel for all proposals for the use of special service subject indicator characters. Such proposals shall meet the following criteria:

- **.1** The full international service must remain unaffected.
- **.2** The special service broadcasts shall be transmitted only when time allows, and with due regard to the necessity for the frequency to remain unused for a high percentage of the time.
- **.3** The special service broadcast shall only be used for its approved purpose.

5.4 B_3B_4 – Message numbering characters (NAVTEX number)

5.4.1 Each message within each subject group is allocated a two-digit sequential serial number beginning at 01 and ending at 99. The B_3B_4 message numbering characters together, are often referred to as the "NAVTEX number".

5.4.2 The NAVTEX number is solely allocated as a component of the NAVTEX message identity and should not be confused with (and bears no correlation to), the series identity and consecutive number of the NAVAREA or Coastal warning contained in the message.

5.4.3 Messages broadcast using NAVTEX number $B_3B_4 = 00$ cannot be rejected and will automatically override any selection of B_1 transmitter identification characters as well as any B_2 subject indicator characters selected on the NAVTEX receiver.

5.4.4 Use of NAVTEX number $B_3B_4 = 00$ must therefore be strictly controlled, since messages carrying it will always be printed or displayed every time they are received. Routine messages and service messages must never be allocated $B_3B_4 = 00$. The correct use of B_2 characters A, B, D and L, will ensure that messages containing safety information will always be printed or displayed on first receipt.

6 Message identity

6.1 The individual NAVTEX message identity is the amalgamation of all four technical characters $B_1B_2B_3B_4$ (transmitter identification character/subject indicator character/message numbering characters).

6.2 When a message is received for the first time by a NAVTEX receiver, the message identity is recorded and stored in the memory for 72 h. This ensures that subsequent transmissions of the same message are not re-printed or repeated in the display, unless they are re-received over 72 h later. In the unlikely event that all 99 NAVTEX numbers for a particular subject group, from a particular transmitter, are in use at the same time, or have been allocated within the past 72 h, an alternative B_2 character must be utilized; for example, $B_2 = L$ has been set aside to be used for additional navigational warnings if all 99 NAVTEX numbers for subject group $B_2 = A$ are in use.

6.3 Each NAVTEX message identity shall be allocated by the relevant NAVTEX Coordinator, who is the authority responsible for the selection of information to be broadcast by each transmitter within each subject group. A single NAVTEX Coordinator may have more than one transmitter under their control. Specific advice on the use of alternative B2 subject indicator characters as mentioned in 6.2 above, can be provided by the IMO NAVTEX Coordinating Panel.

7 Message format

7.1 NAVTEX messages must be composed in accordance with the guidelines contained in the *Joint IMO/IHO/WMO Manual on Maritime Safety Information* and IHO Publication S-53. The format of all messages must be in strict accordance with figure 6. This defines the essential elements of the messages which influence the operation of the receiver. Great care is required to avoid errors of syntax in the groups ZCZC $B_1B_2B_3B_4$ and NNNN as they will cause receivers to operate incorrectly, and may well result in messages not being received.

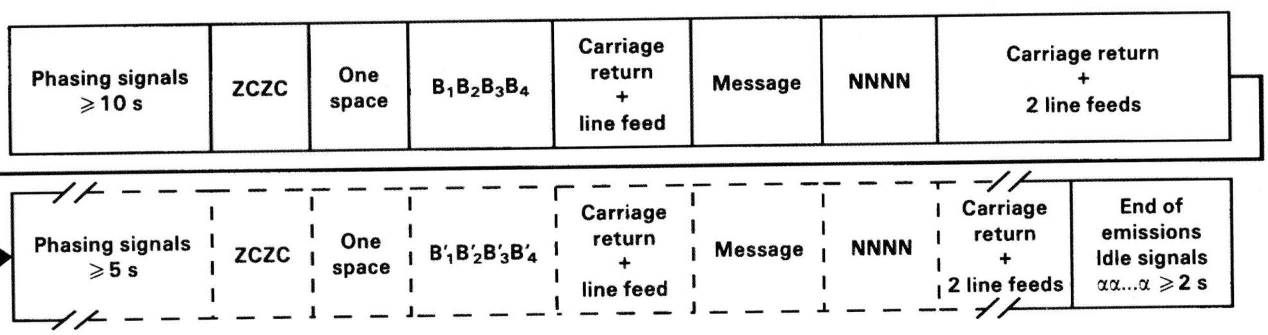

Figure 6 – *Standard format for NAVTEX messages*

7.2 The phasing signal is automatically transmitted by the NAVTEX transmitter at the beginning of each message and is critical to the effective operation of the system. It is this signal which enables a receiver to lock on to a particular station's transmission, providing the frequency is not already in use.

7.3 If another station within transmitting range and with a time slot prior to the station selected overruns its time slot (regardless of the B_1 transmitter identification character in use), its transmission will blank the phasing signal of the subsequent transmitter. It will then seem to the receiver as if the second station is off the air and its broadcast will not be received, possibly denying the user significant safety information. This is the primary reason behind the importance of each station adhering to its allocated time slots. Similarly if the phasing signal for a particular station is too short, some receivers will be unable to lock on to the transmission.

7.4 Basic message elements

Table 3 – *Basic message elements*

Element	Example
Phasing signal	
Start of message group	ZCZC
One space	
NAVTEX message identity	FA01
Carriage return + line feed	
Message content	(Date Time Group – Optional e.g. 040735 UTC OCT 17) NAV I 114/17 ENGLISH CHANNEL. START POINT SOUTHWARD. CHART BA 442 (INT 1701). UNEXPLODED ORDNANCE LOCATED 49-51.97N 003-39.54W AND 49-55.24N 003-40.79W.
End of message instruction	NNNN
Carriage return + two line feeds	
Phasing signal	

7.5 When a message has been received error-free, a record is made by the receiver of the NAVTEX message identity. This unique identifier is used to suppress the printing or display of repeated transmissions of the same message.

7.6 On National NAVTEX services it is important to keep to the same basic message format as that required for the International NAVTEX service. It is also important to ensure that the full broadcast does not overrun the allocated time slot. However, in order to meet national requirements, message content may deviate from the guidelines provided for the International NAVTEX Service if required.

7.7 Examples of navigational warning messages

Note: Further examples are available for consultation in the *Joint IMO/IHO/WMO Manual on Maritime Safety Information*.

```
ZCZC LA18
140356 UTC AUG 17
NORWEGIAN NAV.WARNING 280
CHART 4
AREA OSLOFJORDEN
TORPENE LIGHTBUOY 59-46.1N 010-33.2E UNLIT
NNNN
```

```
ZCZC LA26
250911 UTC JUN 17
DANISH NAVIGATIONAL WARNING NO. 154/17
KATTEGAT, AALBORG BIGHT
LIGHTHOUSE SVITRINGEN RENDE NO.13
56-54.4N 010-30.6E DESTROYED AND MAKES
AN OBSTRUCTION.
DEPTH ABOVE FOUNDATION 1 METRE. THE POS.
IS MARKED AS FOLLOWS:
GREEN LIGHT BUOY Q.G. APPROX 50M SW
YELLOW BUOY APPROX. 50M N
YELLOW BUOY APPROX. 50M ESE
MARINERS ARE ADVISED TO KEEP WELL CLEAR
NNNN
```

```
ZCZC SA38
NAVTEX-HAMBURG (NCC)
131120 UTC SEP 17
NAV WARN NO. 428
TSS TERSCHELLING-GERMAN BIGHT
'TG 2/GW' LIGHTBUOY 53-52N 006-22E
OFF STATION AND DAMAGED.
NNNN
```

```
ZCZC TA93
151530 UTC JAN
OOSTENDERADIO - INFO 17/17
1. OSTEND HARBOUR - WORKING AREA EASTERN
BREAKWATER. ALL SHIPPING (EXCEPT GOVERNMENT
VESSELS AND WORKBOATS INVOLVED IN THIS
PROJECT) FORBIDDEN IN THE WORKING AREA
BOUNDED BY THE FOLLOWING POS:
51-14.278N 002-55.719E
51-14.424N 002-55.696E
51-14.840N 002-55.370E
51-14.579N 002-55.058E
51-14.462N 002-55.186E
51-14.381N 002-55.293E
51-14.253N 002-55.360E
SHIPPING REQUESTED TO PASS WITH REDUCED
SPEED
2. CANCEL INFO 121/09
NNNN
```

```
ZCZC KA79
AVURNAV CHERBOURG 098
DOVER STRAIT TSS
AIS AID TO NAVIGATION
MMSI NUMBER: 992271107
ETABLISHED ON ZC2 BOUY
50-53.6N 001-30.9E (WGS 84)
NNNN
```

```
ZCZC MA99
301435 UTC AUG 17
WZ 972
ENGLAND, EAST COAST.
THAMES ESTUARY.
1. EXPOSED CABLE EXISTS ON SEABED IN
VICINITY OF LINE JOINING:
51-28.7N 000-46.8E
51-29.2N 001-01.7E
51-28.5N 001-09.5E
51-28.8N 001-14.0E
51-28.3N 001-18.6E AND
51-28.7N 001-25.2E.
WIDE BERTH REQUESTED.
2. CANCEL WZ957
NNNN
```

```
ZCZC JA93
101200 UTC SEP
GERMAN NAV WARN 424
WESTERN BALTIC. FEHMARN. PUTTGARDEN.
UNDERWATER OPERATIONS BY 'DEEP DIVER 1/
J8HC7', IN VICINITY OF:
54-32.8N 011-16.9E. GUARD VESSELS
STANDING BY VHF CHANNEL 16. 0.5 NM
BERTH REQUESTED.
NNNN
```

```
ZCZC MA97
291351 UTC AUG
NAVAREA I 238/17
ENGLAND EAST COAST.
THAMES ESTUARY APPROACHES.
CHART BA 1138(INT 1561).
WAVERIDER LIGHT-BUOY AND FOUR GUARD
LIGHT-BUOYS, ALL FL (5) Y.20S,
ESTABLISHED 51-42.5N 001-51.0E.
WIDE BERTH REQUESTED.
NNNN
```

```
ZCZC JA38
051444 UTC AUG
KALININGRAD NAV WARN 097
SOUTHEASTERN BALTIC, KUSHKAYA KOSA
LIGHT LESNOJ 55-01.0N 020-36.8E UNLIT
NNNN
```

7.8 Examples of meteorological messages

Note: Further examples are available for consultation in the *Joint IMO/IHO/WMO Manual on Maritime Safety Information.*

```
OE44
ISSUED BY THE MET OFFICE AT 0620
ON FRIDAY 27 MARCH

GALE WARNINGS: ROCKALL MALIN HEBRIDES SE
ICELAND

THE GENERAL SITUATION AT MIDNIGHT
LOW GERMAN BIGHT 1001 MOV SEAWARDS AND
LOSING ITS IDENTITY. NEW LOW EXP JUST W OF
ROCKALL 989 BY MIDNIGHT TONIGHT

24-HR FCSTS

LUNDY FASTNET
SE VEER SW 5 OR 6, INCR 7, PERHAPS GALE 8
LATER. MOD OR ROUGH, BECMG ROUGH OR VERY
ROUGH LATER. OCCASIONAL RAIN, FOG PATCHES
DEVELOPING. GOOD BECMG POOR OR VERY POOR

IRISH SEA
W 3 OR 4, BACK S 5 OR 6, INCR 7, PERHAPS
GALE 8 LATER.
SLT OR MOD, BECMG MOD OR ROUGH LATER. RAIN
LATER. GOOD, OCNL POOR LATER

...

NNNN
```

```
FICN36 CWIS 310700
ICE NAVTEX FOR IQALUIT VFF AT 0700 UTC
MON 31 AUG 2017.

CUMBERLAND.
WNG: SPECIAL.
1 OI XCPT 8 FYI INCL 3 OI IN THE WRN SECTION.
UNUSUAL PRESENCE OF SEA ICE.

...

NORTHWEST LABRADOR SEA.
WNG: NIL.
BW.

NNNN
```

```
FQCN36 CWNT 251030
NAVTEX FOR IQALUIT VFF AT 5:30 AM EST THU
25 DEC 2016.

VLD 25/11Z-27/05Z,
WND(KT), VIS(NM) ABV 1 UNL IND, FOG IMPL
VIS LESS THAN 1.

SYNOPSIS:
25/12Z INTSF LOW 976 MB OVR SRN QUE. 27/00Z
WKN LOW 965 MB OVR UNGAVA BAY.

RESOLUTION - E:
WNG: STORM / FREEZING SPRAY.
WND: E30. 25/18Z E35. 26/06Z E50. 26/21Z
E30.
SPRAY: 25/11Z-26/11Z MOD-SEV OUT-EDGE
VIS: 26/03Z-27/05Z 0-1 SN.

...

WAVES(M) VLD 25/10Z-27/05Z.

RESOLUTION - E:
2 OUT-EDGE. 26/06Z 4-6. 26/20Z 2

...

NNNN
```

```
FZHW61 PHFO 310955
OFFN10

NAVTEX MARINE FORECAST FOR HAWAIIAN WATERS
NATIONAL WEATHER SERVICE HONOLULU HI
1200 AM HST MON AUG 31 2017

...PLEASE REFER TO COASTAL WATERS FORECASTS
AVAILABLE THROUGH NOAA WEATHER RADIO AND OTHER
MEANS FOR DETAILED COASTAL FORECASTS...

.SYNOPSIS...THE CENTER OF HURRICANE IGNACIO WAS
ABOUT 320 NM EAST OF HILO HAWAII AT 8 PM HST
SUNDAY EVENING...MOVING NORTHWEST AT 8 KT.
IGNACIO IS FORECAST TO MOVE THROUGH THE NORTHERN
OFFSHORE WATERS
THROUGH WEDNESDAY. IGNACIO FORECAST POSITIONS 8
AM HST TODAY 21.4N 151.1W 8 PM HST TONIGHT 22.5N
152.4W 8 AM HST TUESDAY 23.5N 153.8W 8PM HST
TUESDAY 24.5N 155.3W 8 PM HST WEDNESDAY 26.7N
158.7W 8 PM HST THURSDAY 29.0N 162.0W 8 PM HST
FRIDAY 33.0N 163.5W
$$

...
```

```
HAWAIIAN OFFSHORE WATERS

...HURRICANE WARNING...

.REST OF TONIGHT...WINDS 34 KT OR GREATER WITHIN
130 NM OF CENTER.
SEAS 12 FT OR GREATER WITHIN 270 NM OF CENTER.
ELSEWHERE WITHIN
200 NM OF CENTER...WINDS 20 TO 33 KT SEAS 10 TO
15 FT OR MORE.
OTHERWISE... MAINLY NE TO E WINDS 10 TO 20 KT
SEAS 8 TO 10 FT.

ISOLATED THUNDERSTORMS NORTHERN WATERS.

...

NNNN

FQSN40 ESWI 311630

ISSUED 170131
ICE ACCRETION WARNING:

SEVERE ICING IN GULF OF FINLAND.

SEVERE ICING IN NORTHERN BALTIC, SEA OF ÅLAND,
SEA OF ARCHIPELAGO AND GULF OF BOTHNIA.

NNNN
```

```
WWST03 SABM 152100
WEATHER BULLETIN FOR NAVTEX STATIONS - METAREA
6 - JUNE 15, 21:00UTC
NATIONAL WEATHER SERVICE
DATE AND TIME UNIVERSAL TIME COORDINATED - UTC
PRESSURE HPA BEAUFORT SCALE WINDS

STORM WARNING:
WARNING 137: LOW 962HPA AT 54S 72W MOV NE
DEEPENING EXPECTED 52S 52W BY 17/0000 PROVOKES
WINDS FORCE 10 AROUND ITSELF WITH GUST FROM
16/0000

...

NNNN

IB54

JP73 RJTD 270600

IMPORTANT WARNING FOR YOKOHAMA NAVTEX AREA

270600UTC ISSUED AT 270900UTC
PRESSURE GRADIENT IS STEEP
WARNING(NEAR GALE) WESTERN SEA OFF SANRIKU,
NORTHERN SEA OFF KANTO
WARNING(DENSE FOG) EASTERN SEA OFF SANRIKU,
WESTERN SEA OFF SANRIKU
POOR VISIBILITY 0.3 MILES OR LESS IN PLACES
NEXT WARNING WILL BE ISSUED BEFORE 271500UTC

NNNN
```

8 Language and national broadcast options

8.1 International NAVTEX Service messages on 518 kHz shall be broadcast only in English in accordance with resolutions A.706(17), as amended, and A.1051(27), as amended

8.2 There is often a requirement for NAVTEX broadcasts to be made in national languages in addition to English. This shall **only** be achieved by the provision of a National NAVTEX service. National NAVTEX services use frequencies other than 518 kHz, and languages as decided by the Administrations concerned. These National NAVTEX services may be broadcast on 490 kHz or 4209.5 kHz, or on an alternative nationally assigned frequency.

9 Information control

9.1 The time-shared nature of NAVTEX services imposes the need for strict discipline in controlling the information flow of the broadcast. To achieve this, it is necessary to coordinate the messages in each B_2 category at each transmitter. In general, all messages shall be brief and clear and avoid duplication. Strict adherence to relevant guidelines in resolutions A.706(17), as amended, A.1051(27), as amended, and the *Joint IMO/IHO/WMO Manual on Maritime Safety Information* is recommended.

9.2 In addition, certain operating procedures have also been found necessary:

.1 messages in each category should be broadcast in reverse order of receipt by the NAVTEX Coordinator, with the latest being broadcast first; and

.2 cancellation messages should be broadcast once only. The cancelled message should not be transmitted on the broadcast in which its cancellation message appears.

10 Message content

10.1 It is important that national Administrations operating or planning NAVTEX services are clear about what sort of information should be included in the messages.

10.2 The International NAVTEX service should be used for transmitting maritime safety information only and should **not** be used as a medium for providing Notices to Mariners or for broadcasting Local Warnings. NAVTEX is essentially a medium for broadcasting information that is **needed** by ships to safely navigate through the NAVTEX service area of the appropriate NAVTEX station, particularly those ships on coastal passages. More detailed guidance in respect to different classes of messages is given below. Examples of the content and layout of NAVTEX messages are shown in the *Joint IMO/IHO/WMO Manual on Maritime Safety Information*. This publication should be available to all personnel responsible for the drafting of messages to be broadcast by NAVTEX stations.

10.2.1 *Navigational warnings*

.1 coastal warnings and NAVAREA warnings (B_2 = A or L) issued under the guidance of resolution A.706(17), as amended, which would be of concern to ships in the NAVTEX service area allocated to the transmitter should be included in the broadcast. Relevant coastal warnings should normally be repeated at every scheduled transmission for as long as they remain in force; however, if they are readily available to mariners by other official means, for example in Notices to Mariners, then after a period of six weeks they may no longer be broadcast. NAVTEX Coordinators should arrange to receive NAVAREA warnings appropriate to their area for inclusion in their broadcasts. These should be broadcast at least twice each day; to avoid overloading the broadcast time slot, they should normally be scheduled for transmission during slots that do not include weather forecasts (see 12.4);

.2 in-force bulletins (a summary of navigational warnings in force) should normally be broadcast each week; and

.3 local warnings should not be broadcast on International NAVTEX, i.e. information relating to inshore waters, often within the limits of jurisdiction of a harbour or port authority.

10.2.2 *Meteorological warnings and forecasts*

.1 meteorological warnings (B_2 = B), e.g. gale warnings, should be allocated a priority of IMPORTANT (see section 11) and be repeated at subsequent scheduled transmissions for as long as the warning is in force. These messages should contain only the appropriate warnings and should be separate from the weather forecasts;

.2 weather forecasts (B_2 = E) should be broadcast at least twice each day. This service should be carefully coordinated where transmitters are geographically close together;

.3 routine ice reports are normally broadcast on NAVTEX once a day; and

.4 ice accretion warnings (icing warnings) are normally included in gale warnings. If no gale warning is issued, they are to be treated as a meteorological warning (see 10.2.2.1).

10.2.3 *Search and rescue information*

.1 the NAVTEX broadcast is not suitable for distress traffic. Therefore, **only** the initial distress message should be retransmitted on NAVTEX, using B_2 = D, in order to alert mariners to a distress situation, by setting off an audio alarm.

.2 a single authority, which will normally be a Rescue Coordination Centre (RCC), should be designated SAR Coordinator to input information via the NAVTEX Coordinator, for a NAVTEX message. The initial shore to ship distress related message should have previously been broadcast on the appropriate distress frequency prior to any related NAVTEX message being broadcast.

10.2.4 *Piracy warnings*

Piracy warnings should be transmitted using $B_2 = D$, in order to alert mariners by setting off an audio alarm. They should be broadcast immediately on receipt and at subsequent scheduled transmissions.

10.2.5 *Tsunamis and other natural phenomena warnings*

Tsunami, negative tidal surge warnings should be transmitted using $B_2 = D$, in order to alert mariners by setting off an audio alarm. They should be broadcast immediately on receipt and at subsequent scheduled transmissions.

10.2.6 *Pilot and VTS service messages*

Technical subject indicator character, $B_2 = F$, is only to be used for broadcasting temporary alterations, movement or suspension to pilot or VTS services. This category is for the information of all ships and is not to be used for specific instructions to individual ships or pilots.

10.2.7 *No messages on hand*

When there are no NAVTEX messages to be disseminated at a scheduled broadcast time, a brief message should be transmitted to advise the mariner that there is no message traffic on hand. Technical subject indicator character, $B_2 = Z$, is to be used to announce "NO MESSAGES ON HAND".

10.2.8 *Use of abbreviations*

Common examples of abbreviations used in the International NAVTEX service are contained in the *Joint IMO/IHO/WMO Manual on Maritime Safety Information*.

10.2.9 *National NAVTEX services*

Transmissions on 490 kHz or 4209.5 kHz may simply repeat the messages broadcast over the International NAVTEX service but in a national language, or they may be tailored to meet particular national requirements, for example by providing different or additional information to that broadcast on the International NAVTEX service, targeted at recreational vessels or fishing fleets.

11 Message priorities and broadcast procedures in the international NAVTEX service

11.1 Message priorities

11.1.1 The message originator is responsible for assessing the urgency of the information and inserting the appropriate priority marking. One of three message priorities is used to dictate the timing of the first broadcast of a new warning in the NAVTEX service. In descending order of urgency, they are:

VITAL	for immediate broadcast, subject to avoiding interference to ongoing transmissions. Such messages should also be passed to the appropriate NAVAREA Coordinator for possible transmission as a NAVAREA warning via SafetyNET;
IMPORTANT	for broadcast at the next available period when the frequency is unused; and
ROUTINE	for broadcast at the next scheduled transmission.

11.1.2 Both VITAL and IMPORTANT messages should be repeated at each scheduled transmission time slot, if the situation is still valid.

11.1.3 The message priority is a procedural instruction for the NAVTEX Coordinator or the transmitting station and should **not** be included in the message. By selecting the appropriate priority of VITAL, IMPORTANT or ROUTINE at the transmission terminal, the message will be broadcast with the correct priority.

11.1.4 In order to avoid unnecessary disruption to the service, the priority marking VITAL is to be used only in cases of **extreme urgency**, i.e. to relay an initial shore to ship distress related message or acts of piracy warnings, tsunamis and other natural phenomena warnings. In addition, VITAL messages are to be kept as brief as possible and in accordance with the *Joint IMO/IHO/WMO Manual on Maritime Safety Information*. The information provider is responsible for ensuring that the NAVTEX Coordinator is fully and immediately aware when a message should be broadcast with the priority of VITAL.

11.1.5 VITAL messages will normally be broadcast using NAVTEX number B3B4 = 00.

11.2 Broadcast procedures

.1 VITAL priority messages

Messages assessed as VITAL are to be broadcast immediately, subject to avoiding interference to ongoing transmissions. On receipt of a message with a VITAL priority, the NAVTEX Coordinator will commence monitoring the NAVTEX frequency. If the frequency is clear, the VITAL message is to be transmitted immediately. If the frequency is in use, the Coordinator should contact the station which, according to the schedule, will be transmitting during the following time slot and ask it to postpone their transmission start by one minute, to allow a space for the VITAL message. Once the VITAL message has been transmitted, the scheduled station is free to start its routine transmissions;

.2 IMPORTANT priority messages

Messages assessed as IMPORTANT are to be broadcast during the next available period when the NAVTEX frequency is unused. This is to be identified by monitoring the frequency. It is expected that this level of priority will be sufficient for the majority of urgent information; and

.3 ROUTINE priority messages

Messages assessed as ROUTINE, are to be broadcast at the next scheduled transmission time. This level of priority will be appropriate for almost all messages broadcast on NAVTEX and is always to be used unless special circumstances dictate the use of the procedures for an IMPORTANT or VITAL priority message.

11.3 Meteorological NAVTEX messages

The following priorities are to be assigned to meteorological NAVTEX messages:

.1 Tsunami warnings = VITAL

.2 Meteorological warnings = IMPORTANT

.3 Meteorological forecasts = ROUTINE

.4 For other natural phenomena warnings, either VITAL or IMPORTANT may be used.

11.4 National NAVTEX services

The broadcast procedures concerning differing message priorities are the same for both the International and National NAVTEX services.

12 Responsibilities of a NAVTEX Coordinator

12.1 The NAVTEX Coordinator is responsible for the messages transmitted by each station under their control. This responsibility includes checking that the content of each message is in accordance with the *Joint IMO/IHO/WMO Manual on Maritime Safety Information* and also, that it is relevant to the NAVTEX service area of the transmitting station. Thus, a user may choose to accept messages, as appropriate, either from the single transmitter which serves the sea area around their position or from a number of transmitters. Ideally, the user should select the station within whose coverage area their vessel is currently operating and the station into whose coverage area their vessel will transit next.

12.2 The NAVTEX Coordinator must:

.1 act as the central point of contact on matters relating to NAVTEX transmissions for a given transmitter or number of transmitters;

.2 be responsible for continuously ensuring quality control for the operation of the NAVTEX transmitting stations under its jurisdiction. This should be achieved with the cooperation of the information providers to ensure that:

.2.1 messages are always concise and can be transmitted within the designated 10-minute time slots assigned by the IMO NAVTEX Coordinating Panel;

.2.2 MINIMUM power is used to achieve satisfactory range performance; and

.2.3 the coordinated service is operating satisfactorily;

.3 assess all requests for NAVTEX messages immediately upon receipt;

.4 schedule each message for broadcast in accordance with the priority of VITAL, IMPORTANT or ROUTINE;

.5 monitor the International NAVTEX frequency along with any other National frequency used by the transmitters under their jurisdiction in order to ensure that the messages have been correctly broadcast;

.6 monitor the International NAVTEX frequency along with any other National frequency used in order to identify vacant transmission periods required for VITAL or IMPORTANT messages;

.7 pass all information which warrants promulgation outside of their NAVTEX service area directly to the appropriate authority, using the quickest possible means;

.8 allocate a message identity to each message, including the sequential NAVTEX number;

.9 ensure that NAVTEX messages which have been cancelled are removed from the broadcast schedule at the same time as the cancellation message is promulgated;

.10 broadcast in-force bulletins not less than once per week at a regular scheduled time;

.11 promote and oversee the use of established international standards and practices with respect to the format and protocols associated with NAVTEX messages;

.12 maintain records of source data relating to NAVTEX messages in accordance with the requirement of the national Administration;

.13 be aware of the responsibilities of a NAVAREA, Sub-area and National Coordinator contained in resolution A.706(17), as amended, paying particular attention to the specific guidance for the promulgation of internationally coordinated maritime safety information provided therein;

.14 be aware of the responsibilities of a METAREA Coordinator contained in resolution A.1051(27), paying particular attention to the specific guidance for the promulgation of internationally coordinated maritime safety information provided therein; and

.15 take into account the need for contingency planning.

12.3 Management of the service

Data priority:

Most information broadcast on NAVTEX services relates to either navigational warnings or meteorological information. These types of information often originate from different organizations within a country and it is not until they arrive with the NAVTEX Coordinator that an assessment can be made as to whether there is too much information for the relevant broadcast time slot. Each data provider may consider their data to be more important and therefore, require transmission in full. However, the NAVTEX Coordinator needs to control the overall volume of data broadcast and may need to refer back to data providers to prioritize their information and reduce the amount of data to be broadcast. Some NAVTEX Coordinators utilize digital systems which include software that provides a readout of predicted transmission times for data held ready for broadcast. This enables the Coordinator to anticipate any problems and take action before the scheduled broadcast.

Data to meet purely national requirements should not be broadcast on the International NAVTEX service, but should be migrated to a National NAVTEX service (see section 14).

12.4 Balancing the volume of data to be broadcast throughout the daily transmission cycle

12.4.1 For many categories of messages there is no option with regard to when they should be transmitted. However, in order to minimize the risk of over-running the allocated 10-minute time slot, it is possible to balance the overall length of transmissions by broadcasting NAVAREA warnings at different times from weather forecasts and the in-force bulletin. An example of how this may be managed is given below for a station with a B_1 transmitter identification character C:

Time slot	Content
0020–0030	coastal warnings NAVAREA warnings
0420–0430	coastal warnings in-force bulletin
0820–0830	coastal warnings weather forecasts
1220–1230	coastal warnings NAVAREA warnings
1620–1630	coastal warnings ice reports
2020–2030	coastal warnings weather forecasts

13 Best practice for those using the service

13.1 In order to ensure that all necessary maritime safety information has been received, it is recommended that the NAVTEX receiver is switched on at least 12 h before sailing, or preferably left on at all times.

13.2 Logging

13.2.1 The reception of navigational warnings or meteorological information on NAVTEX does not need to be noted in the radio log; the NAVTEX printout (or the non-volatile message memory) satisfies the requirements of regulation 17 of chapter IV of the 1974 SOLAS Convention, as amended.

14 Mutual interference between NAVTEX stations

14.1 The two principal causes of interference are:

.1 transmission overruns; and

.2 excessive power output.

14.2 Although NAVTEX continues to be generally reliable and an effective medium for the promulgation of maritime safety information, the worldwide infrastructure continues to expand and the volume of information that each Administration disseminates through the International NAVTEX service continues to increase. There is a danger that in some geographical areas, without firm management, both the system and system users may become overloaded with information on the single frequency used. This is of particular importance when handling messages of VITAL priority.

14.3 Many stations are filling their allotted 10-minute time slots and an increasing number are over-running. Instances of interference with neighbouring stations, as a result of over-running the time allocation, are also increasing. Where adjacent stations have transmitter identification characters which follow alphabetically (i.e. adjacent time slots), if the first station over-runs, it may mask the phasing signal of the second station such that, to the user, it seems as if the second station is off the air. Maritime safety information from the second station, although broadcast, may not be received by the system users. Over-run is usually caused by one or more of the following which must be avoided preferably by controlling the volume of data broadcast:

.1 a significant increase in safety-critical activity such as cable laying. Navigational warnings promulgating such activity often include numerous waypoints which are listed by Latitude and Longitude;

.2 meteorological information provided in a manner which is not concise and easily assimilated by the system user or for a much wider area than is covered by the NAVTEX station; and

.3 additional information provided for non-SOLAS system users, e.g. longer range weather forecasts for fishing and recreational vessels.

See also section 7.3.

14.4 As the GMDSS spreads to non-SOLAS mariners, their requirements for information are often different from the SOLAS ships and may be determined at a national level. SOLAS ships trading internationally usually pass through the area of coverage of a NAVTEX transmitter in a day; for them a 24-hour weather forecast usually suffices. However, fishing vessels and recreational vessels often remain in the same vicinity for several days and may require much longer range forecasts which take up more transmission time.

14.5 In order to keep the quantity of information that is broadcast on 518 kHz to manageable levels and to reduce avoidable interference on this frequency, Administrations must:

.1 monitor the volume of data broadcast and, together with adjacent Administrations, actively manage the system to ensure that interference caused by over-running allocated time slots is eliminated; and

.2 transmit non-English language broadcasts for SOLAS vessels and broadcasts of information provided specifically for non-SOLAS vessels on 490 kHz or 4209.5 kHz as required. B_1 characters for these frequencies will be allocated by the IMO NAVTEX Coordinating Panel, on request.

14.6 Excessive power output also causes interference between stations with the same B_1 transmitter identification character/time slot, but located in different regions. This has particularly been identified at night, as the number of operational NAVTEX stations increases. Occasionally, this can be caused by atmospheric conditions, but is generally caused by excessive power output from one of the stations. It is recommended that Administrations restrict the power output from their transmitters to that required to cover the designated NAVTEX service area, particularly at night, in order to avoid interference. Experience indicates that the required range of 250 to 400 nautical miles will normally be attained by transmitted power of no more than 1 kW during daylight **with a 60% reduction during night conditions**.

14.7 When interference is detected, particularly when it affects the service to system users, the matter should be addressed immediately. When the interference is with adjacent stations, attempts should be made to resolve the problem locally. Advice may also be sought from the NAVAREA Coordinator. In addition, the IMO NAVTEX Coordinating Panel should be alerted to the problem. When the interference is from a station with the same B_1 character in a different area, the NAVTEX Coordinating Panel must be contacted and they will initiate any necessary investigation/action.

15 Notification of NAVTEX services

15.1 National Administrations should ensure that mariners are informed of the establishment of, and/or changes to, NAVTEX services by inclusion of full details in Notices to Mariners and lists of radio signals. In addition, full details should be forwarded to the appropriate NAVAREA Coordinator, METAREA Coordinator and:

International Maritime Organization
4 Albert Embankment
London SE1 7SR
United Kingdom

Telephone: +44 (0)20 7735 7611
Telefax: +44 (0)20 7587 3210
Email: ncsr@imo.org (in subject line, add: "for Chair, IMO NAVTEX Coordinating Panel")

International Telecommunication Union
Radiocommunication Bureau
Place des Nations
1211 Genève 20
Switzerland

Telephone: +41 22 7305560
Telefax: +41 22 7305785
E-mail: brmail@itu.int

Annex 1
IMO NAVTEX Coordinating Panel terms of reference

1 Terms of reference

.1 advise Administrations planning to implement a NAVTEX service on the frequencies 518 kHz, 490 kHz or 4209.5 kHz, on the operational aspects of the system. In particular, advise on the optimum number of stations, the allocation of transmission identifying characters (B_1) and broadcast message criteria;

.2 coordinate with Administrations over the operational aspects of NAVTEX in the planning stages in order to prevent mutual interference owing to the number of stations, transmitter power, or transmission identifying character assignment;

.3 remain aware of system problems which arise, through reports from sea and correspondence with operational NAVTEX Coordinators;

.4 when problems are identified, liaise with appropriate Administrations involved, NAVAREA Coordinators, METAREA Coordinators, the Sub-Committee on Navigation, Communications and Search and Rescue (NCSR), and the IHO or WMO, as appropriate, to recommend solutions or mitigating measures and, when agreed, coordinate their implementation; and

.5 prepare documentation supporting the system for the NCSR Sub-Committee including those that are needed by Administrations to guide their operations, and those needed to inform the user of the service (mariner, shipowner and operator).

2 Contact addresses

The NAVTEX Coordinating Panel can be contacted at the following addresses:

The Chairman
IMO NAVTEX Coordinating Panel
International Maritime Organization
4 Albert Embankment
London SE1 7SR
United Kingdom

Telephone: +44 (0)20 7735 7611
Telefax: +44 (0)20 7587 3210
Email: ncsr@imo.org (in subject line, add: "for Chair, IMO NAVTEX Coordinating Panel")

3 Panel membership and participation

3.1 The IMO NAVTEX Coordinating Panel is open to membership by all Member Governments and also includes one member nominated by each of the following international organizations:

- **.1** International Maritime Organization (IMO)
- **.2** World Meteorological Organization (WMO)
- **.3** International Hydrographic Organization (IHO)
- **.4** International Mobile Satellite Organization (IMSO)

3.2 The following may be represented as observers on the Panel:

- **.1** IHO World-Wide Navigational Warning Service Sub-Committee
- **.2** International SafetyNET Coordinating Panel
- **.3** Expert Team on Maritime Safety Services (ETMSS) of the WMO/IOC Joint Technical Commission for Oceanography and Marine Meteorology (JCOMM)

3.3 The work of the Panel is conducted mainly by correspondence. Meetings, when appropriate, are announced in advance and normally scheduled to be held in the margins of other IMO or IHO meetings.

Annex 2

Recommendation ITU-R M.540-2[*]

Operational and technical characteristics for an automated direct-printing telegraph system for promulgation of navigational and meteorological warnings and urgent information to ships

The CCIR,[†] (1978–1982–1990)

CONSIDERING

(a) that the availability of navigational and meteorological warnings and urgent information on board ships is of great importance for safety;

(b) that the existing radiocommunication system for promulgation of navigational and meteorological warnings and urgent information to ships can be improved by use of modern techniques;

(c) that IMO has established the following definitions on the promulgation of maritime safety information:

- *NAVTEX* means the system for the broadcast and automatic reception of maritime safety information by means of narrow-band direct-printing telegraphy;
- *international NAVTEX service* means the coordinated broadcast and automatic reception on 518 kHz of maritime safety information by means of narrow-band direct-printing telegraphy using the English language, as set out in the NAVTEX Manual, published by IMO; and
- *national NAVTEX service* means the broadcast and automatic reception of maritime safety information by means of narrow-band direct-printing telegraphy using frequencies and languages as decided by the Administrations concerned;

(d) that the 1988 Amendments to the International Convention for the Safety of Life at Sea, 1974, require that every ship to which the Convention applies shall be provided with a receiver capable of receiving international NAVTEX service broadcasts;

(e) that several countries are operating a coordinated international NAVTEX service, based on narrow-band direct-printing in accordance with article 14A of the Radio Regulations;

(f) that the system should be applicable to the maritime mobile service (both international and national);

(g) that it is desirable that the service fulfils the requirements of all types of ships desiring to use it;

(h) that although each area may need specific guidance, the use of standard technical and operational characteristics would facilitate the extension of the service.

[*] The Director, CCIR, is requested to bring this Recommendation to the attention of the International Maritime Organization (IMO), the International Hydrographic Organization (IHO), the World Meteorological Organization (WMO) and the International Association of Lighthouse Authorities (IALA).

[†] The name "CCIR" was changed to "Radiocommunication Bureau" by the reorganization of the International Telecommunication Union on 1 March 1993.

UNANIMOUSLY RECOMMENDS

1 that the operational characteristics for the promulgation of navigational and meteorological warnings and urgent information using NBDP should be in accordance with annex I; and

2 that the technical characteristics for the promulgation of navigational and meteorological warnings and urgent information using NBDP should be in accordance with annex II.

Annex I

Operational characteristics

1 Narrow-band direct-printing techniques should be used for an automated telegraph system for promulgation of navigational and meteorological warnings and urgent information to ships. Common frequencies for such transmissions should be internationally agreed upon and the frequency 518 kHz has been designated for worldwide use in the international NAVTEX service (see Radio Regulations Nos. 474, 2971B and N2971B).

1.1 For national NAVTEX services Administrations should also utilize the format of this Recommendation on the appropriate frequencies as defined in the Radio Regulations.

2 The radiated power from the coast station transmitter should only be that sufficient to cover the intended service area of that coast station. The range extension occurring during night hours should also be considered.

3 The information transmitted should primarily be of the type used for coastal waters preferably using a single frequency (Resolution No. 324 (Mob-87)).

4 The transmission time allocated to each station should be restricted to that which is adequate for the anticipated messages to be broadcast to the area concerned.

5 Scheduled broadcasts should take place at intervals not exceeding 8 h and be coordinated, to avoid interference with broadcasts from other stations.

6 Message priorities

6.1 Three message priorities are used to dictate the timing of the first broadcast of a new warning in the NAVTEX service. In descending order of urgency they are:

VITAL	for immediate broadcast, subject to avoiding interference to ongoing transmissions;
IMPORTANT	for broadcast at the next available period when the frequency is unused; and
ROUTINE	for broadcast at the next scheduled transmission period.

Note: Both VITAL and IMPORTANT warnings will normally need to be repeated, if still valid, at the next scheduled transmission period.

6.2 In order to avoid unnecessary disruption to the service, the priority marking VITAL is to be used only in cases of extreme urgency, such as some distress alerts. In addition, VITAL messages are to be kept as brief as possible.

6.3 Periods should be scheduled between the regular transmission periods permitting immediate/early transmission of VITAL messages.

6.4 By use of the message serial number 00 in the preamble of a message (see also annex II, paragraph 6) it is possible to override any exclusion of coast stations or of message types which might have been made in the receiving equipment.

7 Initial shore-to-ship distress-related messages should first be broadcast on the appropriate distress frequency by coast stations in whose SAR area distress cases are handled.

8 Participating transmitting stations should be provided with monitoring facilities to enable them to:

- monitor their own transmissions as to signal quality and transmission format; and
- confirm that the channel is not occupied.

9 In case a message is repeated by more than one transmitting station within the same NAVTEX region (e.g. for better coverage) the original preamble B_1–B_4 (see annex II) should be used.

10 In order to avoid overloading of the channel it is desirable to use a single language and where a single language is used it shall be English.

11 Dedicated on-board equipment is recommended.

12 Other operational characteristics and detailed guidance are given in the NAVTEX Manual developed by the International Maritime Organization.

Annex II
Technical characteristics

1 The signals transmitted should be in conformity with the collective B-mode of the direct-printing system specified in Recommendations 476 and 625.

2 The technical format of the transmission should be as follows:

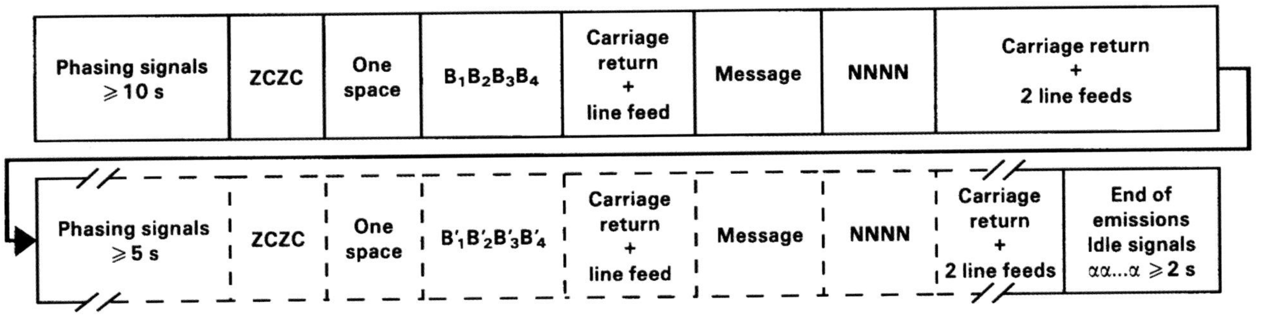

in which

ZCZC defines the end of the phasing period,

the B_1 character is a letter (A–Z)* identifying the transmitter coverage area,

the B_2 character is a letter (A–Z) for each type of message.

2.1 Both the B_1 characters identifying the different transmitter coverage areas and the B_2 characters identifying the different types of messages are defined by IMO and chosen from table 1 of Recommendations 476 and 625, combination numbers 1–26.

2.1.1 Ship equipment should be capable of automatically rejecting unwanted information using character B_1.

2.1.2 Ship equipment should be capable of disabling print-out of selected types of messages using character B_2 with the exception of messages with B_2 characters A, B and D† (see also paragraph 2.1).

2.1.3 If any facility is rejected or disabled in paragraphs 2.1.1 and 2.1.2 above, the extent of any such limitation must be clearly indicated to the user.

2.2 B_3B_4 is a two-character serial number for each B_2, starting with 01 except in special cases where the serial number 00 is used (see paragraph 6 below).

2.3 The characters ZCZC $B_1B_2B_3B_4$ need not be printed.

3 The printer should only be activated if the preamble B_1–B_4 is received without errors.

4 Facilities should be provided to avoid printing of the same message several times on the same ship, when such a message has already been satisfactorily received.

5 The necessary information for the measures under paragraph 4 above should be deduced from the sequence $B_1B_2B_3B_4$ and from the message.

---

* Only letters A–X are used on 518 kHz, 490 kHz and 4209.5 kHz, see table 2 of the NAVTEX Manual.

† B_2 character L (continuation of B_2 subject group A), shall also not be capable of being suppressed (see IEC 61097-6).

6 A message should always be printed if $B_3B_4 = 00$.

7 Extra (redundant) letter and figure shifts should be used in the message to reduce garbling.

8 In case a message is repeated by another transmitting station (e.g. for better coverage) the original preamble B_1–B_4 should be used.

9 The equipment on board ships should be neither unduly complex nor expensive.

10 The transmitter frequency tolerance for the mark and the space signals should be better than ±10 Hz.

Annex 3

Resolution MSC.148(77)
adopted on 3 June 2003

Adoption of the revised performance standards for narrow-band direct-printing telegraph equipment for the reception of navigational and meteorological warnings and urgent information to ships (NAVTEX)

THE MARITIME SAFETY COMMITTEE,

RECALLING Article 28(b) of the Convention on the International Maritime Organization concerning the functions of the Committee,

RECALLING also resolution A.886(21), by which the Assembly resolved that the functions of adopting performance standards for radio and navigational equipment, as well as amendments thereto, shall be performed by the Maritime Safety Committee on behalf of the Organization,

NOTING the carriage requirement in SOLAS chapter IV/7.1.4 for a receiver capable of receiving International NAVTEX narrow-band direct-printing (NBDP) broadcasts for the promulgation of navigational and meteorological warnings to shipping,

NOTING FURTHER the success of the international NAVTEX service in the promulgation of Maritime Safety Information (MSI),

NOTING ALSO with regard to the enhanced storage, processing and display possibilities offered by recent technical advances,

CONSIDERING that further growth in information promulgated to ships will be constrained by the capacity of the International NAVTEX service and the increasing importance of National NAVTEX services,

HAVING CONSIDERED the recommendations on the revision of resolution A.525(13) made by the Sub-Committee on Radiocommunications and Search and Rescue at its seventh session,

1 ADOPTS the revised Recommendation on performance standards for narrow-band direct-printing telegraph equipment for the reception of navigational and meteorological warnings and urgent information to ships (NAVTEX), set out in the annex to the present resolution;

2 RECOMMENDS Governments to ensure that NAVTEX receiver equipment:

(a) if installed on or after 1 July 2005, conforms to performance standards not inferior to those specified in the annex to the present resolution;

(b) if installed before 1 July 2005, conforms to performance standards not inferior to those specified in the annex to resolution A.525(13).

Annex

Revised recommendation on performance standards for narrow-band direct-printing telegraph equipment for the reception of navigational and meteorological warnings and urgent information to ships (NAVTEX)

1 Introduction

1.1 The equipment, in addition to meeting the requirements of the Radio Regulations, the provisions of Recommendation ITU-R M.540, applicable to shipborne equipment and the general requirements set out in resolution A.694(17), should comply with the following performance standards.

2 General

2.1 The equipment should comprise radio receivers, a signal processor and either:

.1 an integrated printing device; or

.2 a dedicated display device,* printer output port and a non-volatile message memory; or

.3 a connection to an integrated navigation system and a non-volatile message memory.

3 Controls and indicators

3.1 Details of the coverage areas and message categories which have been excluded by the operator from reception and/or display should be readily available.

4 Receivers

4.1 The equipment should contain one receiver operating on the frequency prescribed by the Radio Regulations for the International NAVTEX System. The equipment should contain a second receiver capable of working at the same time as the first one on at least two other frequencies recognized for the transmission of NAVTEX information. The first receiver should have priority in the display or printing of received information. Printing or displaying of messages from one receiver should not prevent reception by the other receiver.

4.2 The receiver sensitivity should be such that for a source with an e.m.f. of 2 μV in series with a non-reactive impedance of 50 Ω, the character error rate is below 4%.

5 Display device and printer

5.1 The display device and/or printer should be able to display a minimum of 32 characters per line.

5.2 If a dedicated display device is used, the following requirements should be met:

.1 an indication of newly received unsuppressed messages should be immediately displayed until acknowledged or until 24 h after receipt; and

.2 newly received unsuppressed messages should also be displayed.

5.3 The display device should be able to display at least 16 lines of message text.

5.4 The design and size of the display device should be such that displayed information is easily read under all conditions by observers at normal working distances and viewing angles.

* Where there is no printer, the dedicated display device should be located in the position from which the ship is normally navigated.

5.5 If automatic line feed entails division of a word, this should be indicated in the displayed/printed text.

5.6 When displaying received messages on a display device, a clear indication of the end of a message should be given by automatically adding line feeds after the message or including some other form of delineation. The printer or printer output should automatically insert line feeds after completing print of the received message.

5.7 The equipment should display/print an asterisk if the character is received corrupted.

5.8 Where the printer is not integrated, it should be possible to select the following data to be output to a printer:

.1 all messages as they are received;

.2 all messages stored in the message memory;

.3 all messages received on specified frequencies, from specified locations or having specified message designators;

.4 all messages currently displayed; and

.5 individual messages selected from those appearing on the display.

6 Storage

6.1 Non-volatile message memory

6.1.1 For each receiver fitted it should be possible to record at least 200 messages of average length 500 characters (printable and non-printable) in non-volatile message memory. It should not be possible for the user to erase messages from memory. When the memory is full, the oldest messages should be overwritten by new messages.

6.1.2 The user should be able to tag individual messages for permanent retention. These messages may occupy up to 25% of the available memory and should not be overwritten by new messages. When no longer required, the user should be able to remove the tag on these messages, which may then be overwritten in normal course.

6.2 Message identifications

6.2.1 The equipment should be capable of internally storing at least 200 message identifications for each receiver provided.

6.2.2 After between 60 h and 72 h, a message identification should automatically be erased from the store. If the number of received message identifications exceeds the capacity of the store, the oldest message identification should be erased.

6.2.3 Only message identifications which have been satisfactorily received should be stored; a message is satisfactorily received if the error rate is below 4%.

6.3 Programmable control memories

6.3.1 Information for location (B_1)* and message (B_2)* designators in programmable memories should not be erased by interruptions in the power supply of less than 6 h.

7 Alarms

7.1 The receipt of search and rescue information ($B_2 = D$) should give an alarm at the position from which the ship is normally navigated. It should only be possible to reset this alarm manually.

---

* Refer to Recommendation ITU-R M.540-2.

8 Test facilities

8.1 The equipment should be provided with a facility to test that the radio receiver, the display device/printer and non-volatile message memory are functioning correctly.

9 Interfaces

9.1 The equipment should include at least one interface for the transfer of received data to other navigation or communication equipment.

9.2 All interfaces provided for communication with other navigation or communication equipment should comply with the relevant international standards.*

9.3 If there is no integrated printer, the equipment should include a standard printer interface.

* Refer to publication IEC 61162.

Annex 4

Resolution A.801(19), as amended, annex 4

Criteria for use when providing a NAVTEX service

1 There are two basic areas which must be defined when establishing a NAVTEX service. They are:

Coverage area: An area defined by an arc of a circle having a radius from the transmitter calculated according to the method and criteria given in this annex.

Service area: A unique and precisely defined sea area, wholly contained within the coverage area, for which MSI is provided from a particular NAVTEX transmitter. It is normally defined by a line which takes full account of local propagation conditions and the character and volume of information and maritime traffic patterns in the region.

2 Governments desiring to provide a NAVTEX service should use the following criteria for calculating the coverage area of the NAVTEX transmitter they intend to install, in order to:

- determine the most appropriate location for NAVTEX stations having regard to existing or planned stations;
- avoid interference with existing or planned NAVTEX stations; and
- establish a service area for promulgation to seafarers.

3 The ground-wave coverage may be determined for each coast station by reference to Recommendation ITU-R PN.368-7 and ITU-R Report P.322* for the performance of a system under the following conditions:

Frequency	–	518 kHz
Bandwidth	–	300 Hz
Propagation	–	ground wave
Time of day	–	†
Season	–	†
Transmitter power	–	‡
Antenna efficiency	–	‡
RF S/N in 500 Hz bandwidth	–	8 dB§
Percentage of time	–	90

4 Full coverage of the NAVTEX service area should be verified by field strength measurements.

* Recommendations ITU-R PN.368-7 and ITU-R Report P.322 are superseded by Recommendation ITU-R P.368-9 and Recommendation ITU-R P.372-10, respectively.

† Administrations should determine time periods in accordance with the NAVTEX time transmission table (NAVTEX Manual, table 2) and seasons appropriate to their geographic area based on prevailing noise level.

‡ The range of a NAVTEX transmitter depends on the transmitter power and local propagation conditions. The actual range achieved should be adjusted to the minimum required for adequate reception in the NAVTEX area served, taking into account the needs of ships approaching from other areas. Experience has indicated that the required range of 250 to 400 nautical miles can generally be attained by transmitter power in the range between 100 and 1000 W during daylight with a 60% reduction at night. The receiver characteristics, particularly as regards the bandwidth response, must be compatible with that of the NAVTEX transmitter.

§ Bit error rate 1×10^{-2}.

Annex 5

Procedure for amending the NAVTEX Manual

1 Proposals for amendment or enhancement of the NAVTEX Manual should be submitted to the IMO Maritime Safety Committee through the Sub-Committee on Navigation, Communications and Search and Rescue (NCSR).

2 Amendments to this Manual should normally be approved at intervals of approximately two years or at such longer periods as may be determined by the Maritime Safety Committee. Amendments approved by the Maritime Safety Committee will be notified to all concerned, will provide at least 12 months' notification and will come into force on 1 January of the following year.

3 The agreement of the International Hydrographic Organization and World Meteorological Organization, and the active participation of other bodies, shall be sought according to the nature of the proposed amendments.